L.W.
Butler Community College
901 South Haverhill Road
El Dorado, Kansas 67042-3280

DISCARD

T5-CVO-422

At Issue

Foreign Oil Dependence

Other Books in the At Issue Series

At Issue

Foreign Oil Dependence

Noah Berlatsky, Book Editor

GREENHAVEN PRESS
A part of Gale, Cengage Learning

GALE
CENGAGE Learning·

Farmington Hills, Mich • San Francisco • New York • Waterville, Maine
Meriden, Conn • Mason, Ohio • Chicago

Judy Galens, *Manager, Frontlist Acquisitions*

© 2016 Greenhaven Press, a part of Gale, Cengage Learning.

Gale and Greenhaven Press are registered trademarks used herein under license.

For more information, contact:
Greenhaven Press
27500 Drake Rd.
Farmington Hills, MI 48331-3535
Or you can visit our Internet site at gale.cengage.com

ALL RIGHTS RESERVED.
No part of this work covered by the copyright herein may be reproduced, transmitted, stored, or used in any form or by any means graphic, electronic, or mechanical, including but not limited to photocopying, recording, scanning, digitizing, taping, Web distribution, information networks, or information storage and retrieval systems, except as permitted under Section 107 or 108 of the 1976 United States Copyright Act, without the prior written permission of the publisher.

For product information and technology assistance, contact us at

Gale Customer Support, 1-800-877-4253
For permission to use material from this text or product, submit all requests online at www.cengage.com/permissions.

Further permissions questions can be e-mailed to permissionrequest@cengage.com.

Articles in Greenhaven Press anthologies are often edited for length to meet page requirements. In addition, original titles of these works are changed to clearly present the main thesis and to explicitly indicate the author's opinion. Every effort is made to ensure that Greenhaven Press accurately reflects the original intent of the authors. Every effort has been made to trace the owners of copyrighted material.

Cover photograph copyright © Images.com/Corbis.

LIBRARY OF CONGRESS CATALOGING-IN-PUBLICATION DATA

Foreign oil dependence / Noah Berlatsky, book editor.
    pages cm. — (At issue)
Includes bibliographical references and index.
ISBN 978-0-7377-7368-2 (hardcover) — ISBN 978-0-7377-7369-9 (pbk.)
1. Petroleum industry and trade—Government policy—United States. 2. Petroleum industry and trade--Political aspects--United States. 3. Energy policy--United States. 4. Petroleum conservation--United States. I. Berlatsky, Noah, editor.
HD9566.F673 2016
338.2'72820973--dc23
                                                                2015025024

Printed in Mexico
1 2 3 4 5 6 7 19 18 17 16 15

Contents

Introduction

In 1973, a cartel of Middle Eastern oil producing nations, the Organization of the Petroleum Exporting Countries (OPEC), put an embargo on oil exports to retaliate against American support for Israel. Oil prices spiked and supplies of gasoline dropped, and for the first time Americans were forced to acknowledge just how much they depended on foreign oil, especially from the Middle East. As a result, subsequent presidents and policy makers began to argue for a move toward energy independence. America, they believed, should try to wean itself from foreign oil by consuming less fuel and finding different energy sources.

In response to the 1973 oil shock, the United States instituted a ban on oil exports. The reasoning was that America did not have enough oil to export and that it needed to retain as much of its own supply as possible in order to protect itself from future shocks or embargoes. Some exemptions were made; America, for example, shipped some oil to Canada. However, for the most part, over the last four decades, the United States has not had large oil surpluses, so export has not been much of an issue.

That has changed in recent years, as new technologies like fracking have allowed the United States to access formerly unreachable oil, and has turned the country into an oil and gas producing powerhouse. With oil flowing freely, policy makers and politicians have begun to wonder: should America begin exporting its oil?

Some commentators have insisted that the ban should remain in place. For example, Chris Nelder, writing at *ZDNet*, argues that exporting oil, even given high US production, is not in the country's best interests. He says that allowing American producers to ship oil overseas will effectively increase demand for US oil, which will hurt American consumers. He also contends that shale oil reserves, accessed by fracking, are the last US petroleum resources available. Opening exports will encourage producers to drill out those resources quickly, rather than saving them for a time of need. "Saving some of our remaining oil for a rainy day is a better idea than lifting the crude export ban so we can suck the shale dry tomorrow," Nelder concludes.[1]

Others, however, have argued that lifting the export ban would benefit the United States in numerous ways. Blake Clayton at the Council on Foreign Relations maintains that the limits on oil exports "are already beginning to undermine the efficiency of the US oil economy." Without access to global markets, US oil production is artificially slowed, which deprives the American economy of billions in revenue and thousands of jobs in oil exploration and production. Opening US oil to export would also "demonstrate Washington's commitment to free and fair trade," which would help in US negotiations on trade issues, Clayton argues.[2]

Leon E. Panetta and Stephen J. Hadley at the *Wall Street Journal* contend that ending the export ban would have additional positive effects on American security. They point out that Europe is very reliant on Russian oil, which weakens Europe's ability to respond to Russian aggression in areas like Ukraine. "A recent EU (European Union) 'stress test,'" the authors point out, "showed that a prolonged Russian supply disruption would result in several countries losing 60% of their gas supplies."[3] In addition, revenue from European sales helps Russia fund its military. If America could partially replace

1. Chris Nelder, "Why the US Should Not Export Oil," *ZDNet*, January 26, 2014. www.zdnet.com/article/why-the-us-should-not-export-oil.
2. Blake Clayton, "The Case for Allowing U.S. Crude Oil Exports," Council on Foreign Relations, July 2013. www.cfr.org/oil/case-allowing-us-crude-oil-exports/p31005.
3. Leon E. Panetta and Stephen J. Hadley, "The Oil-Export Ban Harms National Security," *Wall Street Journal*, May 19, 2015. www.wsj.com/articles/the-oil-export-ban-harms-national-security-1432076440.

Russia as a supplier of European fuel, the United States and its allies would be in a much stronger position to negotiate with Russia.

Such arguments appear to have won out with the Barack Obama administration. On July 30, 2014, the tanker *BW Zambesi* left from Galveston, Texas, with a load of crude oil for South Korea. "The 400,000 barrels the tanker carried represented the first unrestricted export of American oil to a country outside of North America in nearly four decades," according to Clifford Krauss at the *New York Times*.[4]

The shipment was enabled through a technicality; the American ban is on crude oil, and the oil on the ship had been lightly refined. Thus, the Obama administration has cagily said there is no real shift in policy. However, Krauss says, "oil company executives and many Wall Street analysts who follow the oil industry interpreted the move as being a fundamental shift in direction."[5] The United States, after forty years, appears to be back in the business of exporting oil. America is not energy independent; it still needs to import certain kinds of oil. But the United States is increasingly becoming, not just an oil importer, but an oil exporter as well.

The remainder of *At Issue: Foreign Oil Dependence* will look at controversial issues surrounding energy independence, including whether energy independence is possible, whether it is desirable, and how renewable energy sources, ethanol, and the Keystone XL Pipeline can, or cannot, reduce America's reliance on foreign oil.

---

4. Clifford Krauss, "In the U.S., a Turning Point in the Flow of Oil," *New York Times*, October 7, 2014. www.nytimes.com/2014/10/08/business/energy-environment /reversing-the-flow-of-oil-.html?_r=0.
5. Ibid.

1

The United States Cannot Attain Energy Independence

Jordan Weissmann

Jordan Weissmann is a senior associate editor at the Atlantic.

The United States and Canada have increased oil and gas production, and they may continue to do so. This would mean that over time the United States would import less from the Middle East. However, it is unlikely that America will ever produce so much oil and gas that it can completely abandon Middle Eastern oil. Even if it could, oil is a worldwide commodity, which means that the price is set by all resources available in the world. Even if the US were not purchasing foreign oil, a drop in supply of that oil would still cause prices to rise worldwide. So increased domestic production of oil is unlikely to make the United States truly independent, as some experts maintain.

American energy independence makes for great political rhetoric. And not much else. We can thank President [Richard] Nixon for the term. During the dark days of the 1973 Arab oil embargo, he publicly vowed to wean the United States off foreign energy sources by the end of the decade, an initiative he dubbed "Project Independence." While things didn't quite pan out the way he imagined, the dream he conjured has lived on with presidents from both parties ever since.

Jordan Weissmann, "The Myth of Energy Independence: Why We Can't Drill Our Way to Oil Autonomy," *Atlantic*, February 9, 2012. Copyright © 2012 Atlantic Monthly Group. All rights reserved. Reproduced with permission.

Politicians Embrace Energy Independence

These days, though, it's not just politicians who are dreaming. Over the last year, it's become respectable—even chic, in a geeky, Washington think-tank sort of way—to suggest that the United States might indeed be close to kicking its foreign energy habit. Take this Bloomberg headline from Monday [February 2012]: "America Gaining Energy Independence." Or this *Financial Times* article from October [2011]: "Pendulum Swings On American Oil Independence." Daniel Yergin, the renowned oil analyst and Pulitzer Prize winner, now argues that the center of world oil production may be moving from the Middle East to the Western hemisphere.

There are plenty of good reasons for the optimism. With the development of its massive shale deposits, the United States has become the world's single largest producer of natural gas. We're so awash in it that domestic prices have plummeted to historic lows. Advances in drilling technology have also made it possible to access hard-to-tap "tight" oil reserves in states such as North Dakota. Analysts believe those fresh crude sources could yield 2.9 million barrels of oil a day by 2020, up from 900,000 today. Meanwhile, cars are getting more efficient, and fuel use has dropped after soaring during the last decade, which frees up more energy production for export. According to Bloomberg, the U.S. is already getting 81% of its energy from domestic sources, the largest share since 1992, and up 10 percentage points since 2005.

Then, there's Canada, which now claims the world's third largest oil reserves thanks to Alberta's petroleum rich tar sands. That's important because Canada and the U.S. are family when it comes to global trade. They're currently our single largest oil supplier. We sell them 75% of their imports and buy 75% of their exports. The National Petroleum Council [NPC], which advises the White House on energy issues, believes that by 2035, the two countries combined could more than double their oil production to 22.5 million barrels a day, enough to satisfy their current total consumption.

If you're not too concerned about how much carbon gets pumped into the atmosphere over the next few decades, these are all great developments. (If you do care about global warming, these are all reasons to have a stiff drink, and perhaps consider moving far from the coasts [since global warming is expected to make sea levels rise]). But even if we're approaching energy independence, the chances of ever actually getting there are rather slim, especially if our economy is still running on oil in 20 years.

American and Candadian crude [oil] would be priced just like everywhere else—based on what the world's highest bidders are willing to pay for it.

Manage Your Expectations

It may theoretically be possible for the U.S. and Canada to more than double our oil output, as the NPC suggests. To put that in perspective, we'd be adding the rough equivalent of another Saudi Arabia to the world oil market. To do it, the countries would have to pretty much tap every resource they have, both onshore and off. Obviously, the old "drill baby drill" crowd would love that approach. But it's still controversial in coastal swing states like Florida. Beyond that, accessing some of the resources would require technology we don't have yet.

In the most likely scenarios, North American oil production will get a big boost in the coming years. It just won't be enough for us to start waving goodbye to OPEC [Organization of Petroleum Exporting Countries]. The U.S. Energy Information Administration forecasts that the American oil production will reach 6.7 million barrels a day by 2020, up from 5.5 million in 2010, then drop back to 6.1 million by 2035.

Canada's National Energy Board foresees future production doubling to 6.0 million barrels per day by that year. So we'd end up with about 12.1 million barrels a day, around two-thirds of what the United States currently chugs through on its own. But let's not be realistic for a moment. Let's assume the U.S. and Canada did manage to drill enough oil that we could tell Saudi Arabia to take its light sweet crude and shove it. What then? Well, we'd still be exposed to all the ugliness of the global oil market. American and Candadian crude would be priced just like everywhere else—based on what the world's highest bidders are willing to pay for it. Americans would continue to feel pain at the pump every time a war broke out in the Middle East or African militants blew up a pipeline.

What We Can Expect

This isn't to say there wouldn't be benefits to greater energy independence. Because natural gas is so difficult to ship, it's not sold on a truly global market, so a big supply at home means cheaper prices. Our abundance of gas has already started luring manufacturers back to the United States in industries, such as chemicals, that rely on it for production. Domestic oil supplies would also help our trade balance. Crude imports account for 44% of the U.S. current account deficit, and buying oil from North Dakota instead of, say, Nigeria would obviously shrink that figure.

But it would still be wise to moderate our hopes, both about North America's ability to drill its way to energy independence, and about what that would even accomplish. Perhaps we can cut down on what we buy from the Middle East. Perhaps we can cut it down significantly. But believing that will save us from the world's problems? That's still just a dream.

2

Reducing US Dependence on Foreign Oil Will Strengthen the Economy

Jason Furman and Gene Sperling

Jason Furman is chairman of the Council of Economic Advisors. Gene Sperling is the director of the National Economic Council.

President Barack Obama has been successful in his efforts to reduce US dependence on foreign oil. This has been achieved mostly through increases in domestic fossil fuel production and in the development of alternative energy sources. These successes benefit the American economy by creating jobs in energy fields. They have also reduced the US trade deficit. Further moves toward energy independence will continue to improve the economy.

Today [August 29, 2013] the Bureau of Economic Analysis revised up its estimate of second quarter [Q2] GDP [gross domestic product] from 1.7 percent to 2.5 percent. This stronger estimate of growth was a result of an upward revision in net exports, with the trade data showing that a key part of the revision is because the trade deficit in petroleum fell to a record low in June. This is yet another reminder that the President's focus on increasing America's energy independence is not just a critical national security strategy, it is also part of an economic plan to create jobs, expand growth and cut the trade deficit.

Jason Furman and Gene Sperling, "Reducing America's Dependence on Foreign Oil as a Strategy to Increase Economic Growth and Reduce Economic Vulnerability," Whitehouse.gov, August 29, 2013.

Reducing Oil Imports

The President [Barack Obama] established a national goal in 2011 to reduce oil imports by one third by 2020 and elevated the goal in 2012 to reduce them by one half by 2020. We are currently on track to meet this ambitious goal if we continue to follow through on the policies that are critical to achieving it.

The oil and gas boom has ... substantially reduced the trade deficit.

There are three basic elements to achieving this goal:

1. Increasing domestic production of oil. Government funded research supplemented private industry's work to develop the technology that sparked the boom in oil and gas production. Crude oil production has grown each year the President has been in office to its highest level in 17 years in 2012. In fact, over the past four years, domestic oil supply growth has accounted for over one-third of global oil production growth.

2. Developing substitutes for oil. This includes almost doubling the production of biofuels since 2007—to a near all-time high—and the substitution as a transportation fuel of oil with natural gas, production of which increased by 25% to an all-time high in 2012.

3. Increase energy efficiency to reduce the use of oil overall. With a combination of the stronger fuel efficiency standards and investments in cutting edge technologies, we currently have the most fuel efficient light-duty vehicle fleet ever, and we are working to increase the efficiency of the medium- and heavy-duty fleet as well.

As a result of these changes, in 2012, net petroleum imports had fallen by one-third since 2008 to the lowest level in

20 years. And imports are continuing to fall this year as well. We will shortly be at the point where domestic crude oil production exceeds imports on a sustained basis for the first time since the early 1990s. The increased domestic supply combined with increased oil efficiency of the economy reduces vulnerability to global supply disruptions and price shocks, enhancing our national security.

Economic Benefits

But among its greatest effects are economic. Every barrel of oil or cubic foot of gas that we produce at home instead of importing from abroad means:

- More jobs. Creates American jobs, adds to our national income, and reduces our trade deficit. Nearly 35,000 jobs have been created over the past four years in oil and gas extraction alone, with more jobs along the crude oil supply chain. North Dakota, for instance, has achieved the lowest unemployment rate in the nation (3.1 percent in June [2013]), while developing into a center of the resurgence of domestic oil production.

- Faster growth. Increasing productivity through new techniques and technologies raises national income and increases growth. And improving the terms-of-trade by reducing America's dependence on foreign oil and increasing our net exports shows up in higher standards of living and also higher growth rates. Most recently, revised net export numbers—including a substantial contribution from petroleum products—played a large role in the upward revision of GDP growth in Q2.

- A lower trade deficit. The oil and gas boom has also substantially reduced the trade deficit. The real (inflation-adjusted) trade deficit in petroleum products fell to a record monthly low in June. The chart below [not shown] shows that through the first six months of

2013, the petroleum deficit is on pace to set a new annual low this year, after adjusting for price changes.

And through June 2013, the petroleum share of the real trade deficit in goods has fallen from over 40 percent in 2009 to 25 percent since then, a pattern that will improve as foreign imports continue to fall and domestic production continues to rise. Economic news like this is just one more reason for us to celebrate the resurgence of domestic oil and gas production.

3

Reducing US Dependence on Foreign Oil Will Not Strengthen the Economy

Sam Mutasem

Sam Mutasem is a senior executive in the power industry.

There is some benefit to the US economy from oil independence in terms of security. However, world oil prices are set by the total supply of oil, whether America imports that oil or not. In addition, oil trade is done in US dollars, which creates a demand for the dollar and thus benefits the US economy. Total energy independence, therefore, would not benefit the US economy, since it would reduce demand for the dollar. Instead, a mix of energy options is the best way to ensure stability and security.

Many in the energy industry, due to varying reasons and drivers, express a great need to reduce or eliminate our dependence on foreign oil as quickly as possible . . . a matter of National Security. I agree the lower the dependence on foreign oil would eliminate many of the economic uncertainties associated with political unrest around the world, disruption of supplies, and the continually increasing demand. These factors impact the cost of living of the American consumer and may squeeze corporate profits.

Global Prices

With the world getting smaller and the economies are interdependent, commodity price in the US will follow the global price. So whether we depend on foreign oil to some extent or

Sam Mutasem, "Dependence on Oil . . . Good or Bad?," energybiz, March 1, 2012. Copyright © 2012 Energy Central. All rights reserved. Reproduced with permission.

eliminate it all together the domestic price of oil will be set by the global market and if the price is up companies will certainly not sell it for less just because we are not importing any oil. As it is, the US imports 20% of its needs from Canada and only 8% from the Middle East. The remainder is produced domestically.

On the other hand, if we drive to reduce the global dependence on oil, until we find an alternative, we will negatively impact the US economy and the US consumer.

Oil, natural gas, and coal will remain the dominant fuels for the foreseeable future because these fuels are abundant and economical.

One fact that most do not realize is that all the oil traded globally is nominated in US dollar. What does that mean? As the demand on oil increases so does the price. As a result the demand on the US dollar will increase and so will the purchasing power of the American Consumer. The Dollar . . . remains King!

Lower Demand for the Dollar

Therefore the drive to reduce dependence on oil may have its benefits, but it will come at a cost that should be mitigated as an integral part of the strategy to reduce dependence on oil. Reducing dependence on oil cannot be approached with a tunnel vision strategy because the lower the dependence on oil the lower the demand on the dollar and the lower the purchasing power of the American consumer. So, what is more . . . a matter of National Security?

In my opinion there is no alternative to a diversified strategy particularly when it comes to energy and natural resources. This includes diversification in the fuel mix we use, the sources of the fuels, and the markets we target. Although we should continue to develop and advance renewable energy, there is no question in my mind that oil, natural gas, and coal will remain the dominant fuels for the foreseeable future because these fuels are abundant and economical. What we need to focus on is making these fuels more environmentally friendly by aggressively investing and developing new technologies to accomplish that. Yes, with such a strategy, there will remain uncertainties that we will have to deal with. However this approach will mitigate our risks and help. Keep the Dollar as King.

The question for those who advocate to eliminate our dependence on oil, how do you propose to maintain the value of the dollar as the demand on the dollar and our purchasing power decrease?

4

US Foreign Oil Dependence
Is a Security Threat

John Miller

John Miller is an energy consultant, researcher, and professional engineer.

There is a serious threat that oil exports from the Middle East to the United States will be cut off by terrorist groups like ISIS (Islamic State of Iraq and Syria) or enemy regimes like Iran. To defend against that threat, the United States needs to increase its energy independence by opening federal lands to fracking oil extraction and by approving the Keystone XL gas pipeline to Canada. These strategies will increase US security and prepare it to go on the offensive against terrorist threats.

President [Barack] Obama has finally recognized the growing threat of ISIS [Islamic State of Iraq and Syria, a terrorist militant group] in Syria & Iraq and to U.S. national security. To address this growing threat the Obama Administration has initiated a new Air Campaign to "degrade and destroy ISIS." And, to possibly avoid the need for future U.S. Armed Forces' "boots on the ground" President Obama is also trying to develop a new Coalition in order to get other Countries to help contain and ultimately destroy ISIS and other developing Middle East terrorist groups. Even though U.S. Energy Security has supposedly been a part of the Obama Administration's "All of the Above" energy policy for years, the recognition of

John Miller, "Growing Middle East Threats to US Energy Security," TheEnergy Collective.com, October 1, 2014. Copyright © 2014 Social Media Today, LLC. All rights reserved. Reproduced with permission.

the potentially growing risks to U.S. petroleum oil imports and Energy Security has been omitted from past and current Administration policies and actions.

Persian Gulf Threat

The Obama Administration apparently has not recognized or acknowledged the potential threats to U.S. Energy Security from growing Middle East oil supply disruption risks. Despite recent increases in domestic crude oil production and somewhat reduced consumption, the U.S. still relies on over 2.0 million barrels per day (MBD) of Persian Gulf crude and petroleum oil imports or almost 12% of total current U.S. petroleum oil consumption.

> *Loss of 20% of total World crude oil supplies from a future Strait of Hormuz shutdown will immediately create an International energy crisis and extremely frantic competition for remaining available oil supplies.*

Due to developing Middle East threats, Persian Gulf imports have the greatest risk of future disruption compared to all other current sources of U.S. petroleum oil imports. Loss of up to 12% U.S. oil imports is almost twice the volume lost during the 1973 Arab OPEC [Organization of Petroleum Exporting Countries] oil embargo.

OPEC Persian Gulf exports must be shipped through the Strait of Hormuz, a narrow waterway that could be readily shutdown if oil shipments were attacked. Just a couple years ago Iran threatened to block all oil shipments through the Strait of Hormuz if Israel or any country were to attack their nuclear development facilities. Due to the less than productive Obama Administration's negotiations to curtail Iran's nuclear weapons development ambitions and the probability of future Israel attacks to protect their country, this risk of shutting down the Strait of Hormuz and Persian Gulf oil shipments is quite real. Remember when Israel took out Iraq's nuclear facility in 1981 in retaliation to a similar national threat. Iran definitely has the military capabilities to readily take out numerous oil tankers that must pass through the Strait of Hormuz that borders their country. Total shutdown of the Strait will block all 17 MBD OPEC Persian Gulf exports or 20% of total current World oil supplies.

If not contained, the growing ISIS and other terrorist groups such as Al Qaeda could expand from Syria and Iraq possibly into other Middle East Countries including Kuwait, Saudi Arabia, Qatar and United Arab Emirates. ISIS, Al Qaeda and other terrorist groups could also partner with Iran behind the scenes, and coordinate attacks against other OPEC Persian Gulf oil facilities and shipping infrastructures in retaliation of U.S. or Coalition attacks in Syria or Iraq, or possible further deterioration of the Iranian nuclear negotiations.

Major Disruptions

The shutdown of the Strait of Hormuz will not only disrupt all U.S. Persian Gulf imports, but will also impact the availability and cost of nearly all other oil imports from outside North America. Loss of 20% of total World crude oil supplies from a future Strait of Hormuz shutdown will immediately create an International energy crisis and extremely frantic competition for remaining available oil supplies. Strong competition for the balance of world oil supplies will at minimum substantially increase crude oil market prices very rapidly. Since the U.S. currently imports over 3.0 MBD of oil from non-Persian Gulf countries outside North America, the limited availability and substantially higher cost of over 15% of total current U.S. crude and petroleum oil supplies from Countries other than Canada and Mexico, will further decrease Energy Security and negatively impact the overall economy beyond just loss of 2.0 MBD of Persian Gulf imports.

To help mitigate the impacts of possible future U.S. oil supplies disruptions the Federal Government built the Strategic Petroleum Reserve (SPR) following the 1973 Arab OPEC oil embargo. In the event of lost imported oil, the SPR can replace up to 4.4 MBD directly to the Gulf Coast and Mid Continent via existing pipeline, rail and inland marine infrastructures. SPR transfers to the East and West Coast will be more constrained due to limited "Jones Act" shipping. The 1920 Jones Act requires U.S. port-to-port shipments normally be made in U.S. built, crewed and operated tanker ships. The West Coast SPR shipments are further constrained by the required Panama Canal shipment routing. If or when the Strait of Hormuz is shutdown, the oil shortages on the West Coast, and East Coast to a lesser degree, could approach the historic energy crises levels experienced following the 1973 Arab OPEC embargo and the 1979 Iranian Revolution.

The large threats to Persian Gulf oil shipments and U.S. Energy Security need to be immediately addressed to avoid a future energy crisis which could cause another "Great Recession." The Obama Administration needs to acknowledge these risks and implement new policy changes needed to mitigate the growing threats to U.S. Energy Security. Since the progress in reducing U.S. petroleum consumption has slowed significantly as the economy recovers more fully from the Great Recession, this possible solution to significantly reducing future oil imports will likely take 10–20 years to significantly reduce or eliminate Persian Gulf imports. The most timely and effective strategy to reducing or possibly eliminating the U.S. need for Persian Gulf oil imports is further increasing North America crude oil supplies. This will involve major changes to the Obama Administration's existing "All the Above" energy policy and expediting the development of all forms of the most secure sources of required petroleum supplies.

More Federal Fracking

Clearly the single largest impact on U.S. reduced oil imports recently has been due to increased domestic crude oil production from newly developing "hydraulic fracturing" technology. As a result of this new technology Oil Companies have been able to produce large and growing amounts of unconventional "tight oil" from the huge shale reserves around the Continental U.S. Even though the Obama Administration tries to claim credit for this increase in U.S. domestic production, the reality is that with very few exceptions, the vast majority of increased "tight oil" production has come from State and Private land leases; and not leased Federal properties. To further increase domestic production of unconventional and conventional crude oil the Obama Administration needs to immediately increase the access and number of leases issued for Federal on- and off-shore oil reserves. This policy change will accelerate future domestic crude oil production and more rapidly reduce currently needed and high risk Persian Gulf oil imports.

Cost effectively reducing the need for petroleum fuels will be critical for reliably sustaining and growing the future U.S. economy.

The next major policy change the Obama Administration needs to make is immediately approving the Keystone XL pipeline. This action will give the U.S. very quick access to the most secure source of oil imports from the U.S.'s largest Trade Partner and important Ally; Canada. It's time to stop the seemingly endless delays in approving the cross-border section of the Keystone XL pipeline project. After six years no significant or un-resolvable environmental issues have been identified and approving the Keystone XL will lead to the quickest and largest reduction in Persian Gulf imports, or up to 0.7 MBD within a couple years.

Besides more rapidly opening up Federal lands/waters to significantly more oil production development, the Obama Administration needs to properly manage the exports of light tight (crude) oil or condensates, and LNG [Liquified Natural

Gas]. Keeping light tight oil (LTO) within the U.S. will encourage Domestic Refiners to make the investments needed to efficiently process increased volumes of LTO in the future. Keeping LNG within the U.S. will make the economics of developing alternative natural gas fueled vehicles attractive enough to displace increased volumes of diesel and gasoline petroleum motor fuels in the future. In addition to better managing current and potential future oil & gas exports, the Administration needs to persuade Oil Companies and Traders to reduce the level of imports purchased from Persian Gulf Countries and substitute safer, more secure oil imports from other Countries around the world.

The Obama Administration definitely needs to continue its support for improved energy efficiency of all the different technologies that use petroleum as a primary fuel. This includes reduced Transportation petroleum motor fuels consumption and fuels switching from petroleum-to-natural gas in the Industrial, Commercial, Residential and Power Sectors. An effective "All of the Above" Energy Security policy needs to increase domestic production and reduce higher risk imports until actual reductions in required petroleum consumption can be achieved by fuels switching to cleaner, renewable fuels including electric vehicles, and increased fuel efficiency technologies. Cost effectively reducing the need for petroleum fuels will be critical for reliably sustaining and growing the future U.S. economy.

Implementing an effective new U.S. Energy Security policy that will reduce the need for Middle East OPEC Persian Gulf imports could also become the basis for a more effective strategy to degrade and destroy ISIS and other terrorist groups in the Middle East. By substantially reducing and possibly eliminating the purchase of OPEC Middle East crude and petroleum oil imports in the near future, this strategy would also become the basis for a new "Trade War" to defund current and future terrorist organizations. Such a strategy would help reduce the potential financial support of growing terrorist threats and could be much more effective than current Air Force or limited future "Boots on the Ground" campaigns.

5

Energy Independence Will Not Break OPEC Influence

Gal Luft and Anne Korin

Gal Luft and Anne Korin are codirectors of the Institute for the Analysis of Global Security, a nonprofit think tank located in Potomac, Maryland.

Producing more oil and achieving energy independence will not help the United States break free of the influence of the Middle Eastern oil cartel (Organization of Petroleum Exporting Countries, also known as OPEC). Oil markets are global; as long as OPEC can slow its own rate of production, it can increase world oil prices, no matter how much oil the United States produces. If the United States wants to break OPEC's influence, it needs to find another commodity that can be used to power cars and other transportation. If more cars ran on natural gas, for instance, the price of gas would increase and oil prices would fall.

The first U.S. energy secretary, James Schlesinger, observed in 1977 that when it comes to energy, the United States has "only two modes—complacency and panic." Today [2013], with the country in the middle of an oil and gas boom that could one day crown it the world's largest oil producer, the pendulum has swung toward complacency. But 40 years ago this week, panic ruled the day, as petroleum prices quadrupled

Gal Luft and Anne Korin, "The Myth of US Energy Dependence," *Foreign Affairs*, October 15, 2013. Copyright © 2013 Foreign Affairs. All rights reserved. Reproduced with permission.

in a matter of months and Americans endured a traumatic gasoline shortage, waiting for hours in long lines only to be greeted by signs reading "Sorry, no gas."

The Oil Embargo

The cause of these ills, Americans explained to themselves, was the Arab oil embargo—the decision by Iran and the Arab members of the Organization of Petroleum Exporting Countries (OPEC) to cut off oil exports to the United States and its allies as punishment for their support of Israel in the 1973 Yom Kippur War. And the lessons they drew were far-reaching. The fear that, at any given moment, the United States' oil supply could be interrupted by a foreign country convinced Washington that its entire approach to energy security should center on one goal: reducing oil imports from that volatile region.

Americans have been led to believe that the vulnerabilities associated with oil dependence would be alleviated if only oil imports decreased.

But Americans were wrong on both counts. The embargo itself was not the root cause of the energy crisis. Contrary to popular belief, the United States has never really been dependent on the Middle East for its supply of oil—today only nine percent of the U.S. oil supply comes from the region. At no point in history did that figure surpass 15 percent. Rather, the crux of the United States' energy vulnerability was its inability to keep the price of oil under control, given the Arab oil kingdoms' stranglehold on the global petroleum supply. Nonetheless, for the last four decades, Washington's energy policy has been based on the faulty conclusion that the country could solve all its energy woes by reducing its reliance on Middle Eastern oil. Where did this conclusion come from? By the time the six-month embargo was lifted, in March 1974, the global economy lay in ruins. In the United States, unemployment had doubled and GNP [gross national product] had fallen by six percent. Europe and Japan had fared no better, and struggling, newly created countries in Asia and Africa took the worst hits. Countries completely dependent on energy imports found themselves heavily in debt, and millions of unemployed poor had to migrate from the cities back to their villages.

The crisis also dealt a blow to American prestige. At the height of the Cold War, the United States essentially proved that without oil it was a paper tiger. The worried secretary of state, Henry Kissinger, indicated that the United States was prepared to send military forces to the Persian Gulf to take over whatever country was needed to keep the oil flowing. Since 1973, the United States has sent forces to the Middle East time and again in the name of energy security. Moreover, the embargo created a deep sense of vulnerability from which the United States has never recovered. The country has been portrayed that way by its own leaders: in 2006, Senator Joseph Lieberman called it "a pitiful giant, like Gulliver in Lilliput, tied down and subject to the whim of smaller nations."

Energy Independence Won't Work

The only proper response, it seemed, was to stop importing so much Middle Eastern oil. Every U.S. president since the embargo, from Richard Nixon to Barack Obama, has sought the elusive goal of "energy independence," either by increasing domestic oil supply (Republicans) or by constraining demand through a gasoline tax and improving the standards for cars' fuel efficiency (Democrats). Americans have been led to believe that the vulnerabilities associated with oil dependence would be alleviated if only oil imports decreased. Furthermore, they have been promised that import reduction would yield lower crude prices and thus lower prices at the pump.

Those assertions were wrong 40 years ago and they are even further off the mark today. The long race for energy self-sufficiency reflects a systematic failure to grasp the meaning of the events of 1973—specifically the exact role that OPEC played during this episode and over the subsequent four decades. It is time to take a fresh look at those events, to rethink the U.S. national fixation with energy self-sufficiency, and to focus on solutions that actually have a chance of getting the United States—not to mention the rest of the world—out of the mire.

The Embargo Syndrome

The first clue that the oil embargo did not cause the United States' energy woes is that the real (inflation-adjusted) price of oil barely dipped when the embargo ended and did not again hit the pre-embargo lows until the late 1990s. The fundamental driver of the rise in oil prices was rather a structural shift in the oil market, which transformed it from a buyer's to a seller's market. From the mid-1940s to the 1970s, oil markets were dominated by the so-called Seven Sisters, investor-owned Western oil companies that controlled the global petroleum industry. They were replaced by a cartel of 12 governments.

In the absence of adequate uses for their oil wealth, the Arabs would likely conclude that oil in the ground was just as good as money in the bank and that they should produce less rather than more.

OPEC was initially created in 1960 by five member states that were frustrated. They felt that they earned too low a share of oil revenues, and they were irritated by oil import quotas set by the United States in 1959 that lowered oil prices outside North America while keeping them high for the benefit of domestic drillers. Moreover, they were intent on changing the balance of power between themselves and the investor-owned oil companies. But the organization did not garner real power until the following decade, after the United States became a net energy importer in 1971.

OPEC's founders understood that by consolidating control over a large portion of the world's oil reserves and colluding to suppress oil production, they could drive prices up to a level more to their liking. In the three years prior to the embargo, OPEC members worked hard and fast to seize control over the international oil market. They taxed and nationalized their oil assets and implemented arbitrary production cuts and sharp increases in prices to offset the loss of their income caused by the decline in the value of the dollar. These measures effectively doubled the price of crude oil between 1970 and 1973.

A sense of impending doom was in the air. In an influential April 1973 *Foreign Affairs* article, James Akins, a White House oil expert who was appointed U.S. ambassador to Saudi Arabia a month before the embargo began, predicted an oil crisis. In the absence of adequate uses for their oil wealth, the Arabs would likely conclude that oil in the ground was just as good as money in the bank and that they should produce less rather than more, despite the blistering growth in global demand. Before the Arabs and the Israelis exchanged a single bullet, OPEC was already working hard to drive oil prices up.

Then came the actual embargo. On October 19, 1973, the Arab members of OPEC and Iran decided to stop sending oil to the U.S. market as a punishment for President Richard Nixon's appeal to Congress to appropriate $2.2 billion in emergency aid for Israel. What mattered most, however, was not the decision to cut off exports but the cartel's throttling down of oil production.

The oil market is like a pool into which producers pour oil and from which consumers take it out. It does not matter so much who purchases which oil from whom. If the embargo simply consisted of a ban on exports to particular countries, it would not have had much of an impact on prices, since those countries would have purchased from an alternate supplier and OPEC's oil would have flowed elsewhere. Yet a reduction of supply in the face of the same level of demand was guaranteed to drive prices up globally—for everyone—not just to those countries targeted by the embargo. What really happened was that key members of OPEC took advantage of geopolitical events to shift to a lower level of supply and to send prices up to what they perceived as a more just level. In total, five million barrels per day were withdrawn from the market, and OPEC's posted price of crude doubled yet again, from $5.12 to $11.65 per barrel.

US Policy Caused the Shortage

Meanwhile, in the United States, another policy had already set the stage for the snaking gas lines and desperate drivers. The Economic Stabilization Act of 1970 gave the president control of wages, rents, and prices across the American marketplace, including the price of fuel. Whereas in 1970, the Mandatory Oil Import Quota Program had kept U.S. oil prices about 2.5 times higher than global prices, and politicians said nothing, the major price spike following OPEC's 1973 production cuts sent the political and regulatory machinery into a spin. Politically unable to unwind fuel price controls and let the price of gasoline go up in sync with rising global oil prices, the U.S. government had made selling fuel in the United States a losing proposition for some refiners. This caused a reduction in domestic fuel supply. Demand did not drop because the government prevented prices from rising in a way that reflected market realities. The result was shortages at the pump, a spread of panic and uncertainty among buyers, and a doubling down by the government: the Emergency Petroleum Allocation Act, passed in November 1973, enabled the administration to embark on Soviet-style allocation and rationing of petroleum products.

Energy security is traditionally defined as the availability of sufficient supply at affordable prices. The collective memory of the embargo and the U.S. response to it were mostly shaped by the events that were perceived to affect availability—the embargo and the gas lines—rather than OPEC's change of the supply-demand balance, which for decades has affected the affordability side of the ledger. Nixon's response to the crisis, Project Independence, [was] aimed at achieving energy self-sufficiency for the United States by 1980, but it ignored the real story: the cartelization of the world's most important commodity and the new balance of power that had been established between consumers and producers.

The Anatomy of a Cartel

Analysts tend to discount OPEC's role in the modern energy market, deriding it as a dysfunctional and irrelevant group that long ago lost its sway in setting oil prices. Watching OPEC's conduct on a week-by-week basis, especially the internal disputes among its members, that conclusion seems plausible. But looking at the cartel's overall performance since 1973, one can appreciate the precision of Akins' observation that for OPEC, oil in the ground is as good as oil in the bank. In the past 40 years, the world's population has grown from four billion to seven billion, the number of vehicles in the world has quadrupled, and the Chinese economy has risen from its slumber. All these trends have caused global oil demand to spike from 55 million barrels a day in 1973 to 88 million barrels a day today. Although the United States and other non-OPEC producers have been increasing their production, OPEC, which holds some three-quarters of the world's conventional oil reserves and has the lowest per-barrel production costs in the world, produces today the exact amount of oil it did four decades ago: 30 million barrels a day, accounting for about a third of global supply. In other words, OPEC deliberately produces much less oil than its reported reserves would allow in order to keep prices higher than they would otherwise be. If investor-owned oil companies such as Exxon, BP, Shell, and Chevron were sitting on top of three-quarters of the world's conventional oil reserves, they would be supplying something around three-quarters of the world's oil. And if not, they'd be slapped with an antitrust lawsuit. Antitrust lawsuits, however, don't work against sovereign governments, and sovereign governments are what constitute OPEC.

> *As long as oil is essentially the world's sole transport fuel, neither expanded domestic oil production nor improvements in the efficiency of cars will change this reality.*

At the same time, the Arab OPEC members today face growing budgetary obligations as a result of the Arab Spring's unrest.[1] They need to maintain an oil price high enough to ensure that they will have enough money to distribute to keep the masses from storming the palaces. To make matters worse, Persian Gulf countries are also among the world's fastest-growing oil consumers—Saudi Arabia, for example, is the sixth-largest oil consumer, using more oil than Germany, South Korea, or Canada—which means they have less oil for export as their domestic demand grows. What OPEC terms the "fair price" or the "reasonable price" of oil—which in practice means whatever price its members require to balance their national budgets—will remain high in order to ensure political stability. "In 1997, I thought $20 was reasonable. In 2006, I thought $27 was reasonable," the Saudi oil minister, Ali al Naimi, said in March this year [2013]. "Now, it is around $100 . . . and I say again, 'It is reasonable.'"

The financial obligations of OPEC members are likely to continue to inflate, and so OPEC's response to the oil boom

1. The Arab Spring refers to uprisings and revolutions in Middle Eastern nations beginning in late 2010.

L.W. Nixon Library
Butler Community College
901 South Haverhill Road
El Dorado, Kansas 67042-3280

in the Western Hemisphere, which has the potential to drive down energy prices, will need to be corresponding to cuts in production. This has been the cartel's *modus operandi* since its inception. When non-OPEC producers such as the United States or Norway increase their production, OPEC can respond by decreasing supply accordingly, keeping the overall amount of oil in the market the same. With annual revenues exceeding $1 trillion, OPEC members seem unconcerned by the pain they have inflicted on the global economy, not to mention the world's poorest nations, with oil's meteoric price rises.

Put simply, what Americans import from the Persian Gulf is not so much the actual black liquid as its price. As long as oil is essentially the world's sole transport fuel, neither expanded domestic oil production nor improvements in the efficiency of cars will change this reality. Such remedies may have a positive impact on our trade balance and the environment, but they will have little bearing on the economic burden of importing oil or the price that consumers will face.

In every sector in which oil has faced competition from substitute commodities, it has lost market share due to its high price.

Compete, Baby, Compete

Half a century of a global transportation sector dominated by OPEC has led us to accept the cartel's price-over-volume strategy as a *fait accompli*. We shouldn't. In a modern global economy defined by free trade, open markets, and antitrust laws, no cartel should be allowed to dominate any commodity, not least the most strategic one of all. That most OPEC members adhere to the World Trade Organization's obligations and that one of them, Qatar, is even home to the WTO's current trade negotiations round only highlight the inconsistency.

What can be done? A breakup of OPEC is unlikely, since all its members are in the same boat, and holding on to the cartel is their only way of remaining economically viable and maintaining domestic stability.

But the United States now has a unique opportunity to stabilize oil prices—and to do so in a fairly short period. To see how, one can look to salt, the commodity that for most of human history held the same strategic importance that oil holds today. Just as oil has a virtual monopoly over transportation fuel, salt was for centuries the only means of food preservation. As a result, the pursuit of salt was a matter of war and peace and a source of real conflict. This grip was broken with the invention of competing methods of food preservation, such as canning and refrigeration. Salt's strategic importance was eliminated not because countries stopped importing or using salt—in fact, the United States imports and uses more salt than ever before—but because there were other options. Oil's strategic importance can be similarly reduced.

All that is needed to enable a car to run on methanol are a fuel sensor and a corrosion-resistant fuel line.

In every sector in which oil has faced competition from substitute commodities, it has lost market share due to its high price. For example, until 1973, most industrialized countries, and certainly the developing ones, used oil to generate electricity. As much as a quarter of U.S. electricity and 70 percent of Japan's was generated from petroleum. But the emergence of civilian nuclear power in the 1970s, followed by the increase in the use of natural gas for power generation, effectively pushed oil out of the electricity mix. Today, in the United States, only one percent of electricity comes from petroleum, and only one percent of U.S. petroleum demand is due to electricity generation. Despite popular claims that drawing electricity from wind turbines, solar panels, and nuclear power would reduce the world's oil dependency, today in most countries, including China, the electricity sector is decoupled from oil.

Such a transformation has yet to happen in the transportation sector, which remains as dominated by oil as it was four decades ago. Other energy commodities, such as natural gas and coal and the fuels that can be made from them, are much less costly than oil and oil-based fuels. Yet so long as cars are made and certified to run on nothing but petroleum fuels, oil will continue to be an uncontested master and will not face competition at the pump. Should cars be open to a variety of fuels, however, oil would be forced to compete for market share, and this competition would force oil prices down even as it drove the price of other energy commodities higher. The ubiquity of the petroleum-only vehicle, despite the existence of cheaper non-petroleum fuel, is partially a result of the challenge of coordinating between fuel stations, fuel makers, and vehicle manufacturers, and partially due to entrenched regulatory advantages petroleum has accumulated over the years. These can and must change.

Pitting Natural Gas Against Oil

The recent proliferation of fracking and horizontal drilling technologies has unleashed such substantial quantities of tight oil and natural gas in North America that it has become a cliché to proclaim an "energy revolution." But if this development is to have any real and lasting impact on the security of the global oil supply, it will stem from unconventional natural gas production rather than from unconventional oil. Increases in domestic oil production are, after all, trivial for OPEC to counter. Low-cost natural gas is another story.

At current oil and natural gas prices, oil costs five times more than natural gas on an energy equivalent basis. But despite its low cost, less than one percent of U.S. natural gas is used to fuel automobiles. There are a number of paths to making use of natural gas in transportation. Some would allow for cheap fuel but would increase the cost of vehicles; others would be able to keep down the cost of both. For example, using compressed natural gas to power vehicles, while quite cheap on the fuel side, would make cars more expensive, since a gaseous fuel under pressure requires a much more expensive fuel tank than a liquid fuel for safety reasons. Electricity generated from natural gas could power plug-in hybrids and electric vehicles—also somewhat costly on the vehicle side and quite cheap on the fuel side. Natural gas could also be converted to liquid fuels such as gasoline, ethanol, and methanol, all of which could be used by engines capable of working on any blend of gasoline and alcohol. This last option would add roughly $100 to the cost of a vehicle.

Methanol offers a particularly appealing alternative because of its affordability (it sells today for a dollar less than a gallon of gasoline on an energy-equivalent basis), scalability, and the very low cost of enabling vehicles to use it. All that is needed to enable a car to run on methanol are a fuel sensor and a corrosion-resistant fuel line. And in fact, China has opened its transportation fuel market to methanol and has become the world's largest producer and user of the fuel, which in China is primarily made from coal. The fuel's economics are so attractive that illegal blending of methanol and gasoline is rampant.

Opening vehicles to fuel competition would pit cheap and abundant commodities such as natural gas and coal against one whose supply is chronically constrained by a cartel and whose price is consequently inflated. The subsequent increase in production capacity of non-petroleum fuels, and the ability to shift on the fly among different fuel sources at the pump depending on their per-mile pricing, would finally allow market competition to drive down the price of oil.

The realities of geology and the comparative marginal cost of production in different regions make it extremely unlikely that OPEC can be knocked out of its monopolist position in the global oil market. But fuel-competitive vehicles would make the cartel just another purveyor of commodities when it comes to the transportation fuel market. For this to happen, the United States first needs to realize that its current approach might bring oil self-sufficiency, but it will get the country nowhere near energy security. True energy security would not require the United States to shield itself from the rest of the world. Rather, it would require the United States to apply to the transportation fuel sector the economic principles it has always cherished: consumer choice, open markets, and vigorous competition.

6

The Keystone XL Pipeline Is Not a Good Energy Choice

Brad Wieners

Brad Wieners is an executive editor for Bloomberg Business-week.

The Keystone XL Pipeline will not create many jobs. It is unnecessary in terms of energy independence, since the United States has been developing new resources and increasing energy production dramatically in the last few years. It is an environmental disaster in terms of climate change and adding carbon to the atmosphere. There is no good reason for President Barack Obama to approve the project. His failure to kill it is a mystery.

Give me one good reason [Barack] Obama should Approve Keystone XL.

Really, there could be two:

1. President Obama, Secretary of State John Kerry, and economic growth-focused Washington want China as America's new BFF [best friend forever] and plan to let Beijing [capital of China] know by offering up an energy supply from our friends to the North.

2. Obama, Kerry, and Canadian Prime Minister Stephen Harper have worked out a *quid pro quo*. The Yanks will accept a pipe carrying toxic sludge through America's bread basket so long as Canada takes over counterterror-

Brad Wieners, "Give Me One Good Reason Obama Should Approve Keystone XL," *Bloomberg Business*, March 7, 2014. Copyright © 2014 Bloomberg Business. All rights reserved. Reproduced with permission.

ism in Afghanistan, sends peacekeepers to Ukraine, and Harper himself places [Canadian popstar] Justin Bieber under house arrest so he can't tour in the lower 48.

Some American teens might not find that last measure in the national interest, but some version of these realpolitik rationales—overture to China, huge favor to Harper—are about the only ones left to explain why Obama hasn't killed the proposed 875-mile final leg of pipeline from Alberta [Canada] to the Gulf of Mexico.

Not Enough Jobs

A number of strong arguments appeared to be in favor of Keystone XL when it first became a national story, beginning with jobs. Several U.S. representatives and senators testified that the pipeline would yield 20,000, 40,000, or even 100,000 new jobs. The recession made those prospects extra compelling. Turns out they were extra optimistic, too. Now we know the pipeline might generate about 3,900 temporary (two-year) construction jobs and about 50 permanent ones. (Should we really be surprised? The whole point of a pipeline is that it's automated.)

The other big case for Keystone—also given full voice by pols who received campaign help from oil and gas lobbies—was the chance to rely on a friendly neighbor for oil rather than on an unstable Middle Eastern regime. But now, due in part to fracking and the Bakken reserve in North Dakota, U.S. oil inventories are at a 21-year high; a glut of unrefined oil is sitting in Cushing, Okla., and the U.S. is expected to become the world's leading oil producer next year. Moreover, the sweet crude pouring out of the Bakken is of far finer quality than bitumen, the sour, thick oil sands extraction that is effectively steamed out of the soil beneath Alberta's former boreal forest. What's more, Keystone XL isn't really designed to serve the U.S.; it's meant to get Alberta's tar sands to Texas refineries and ready for export. The Keystone XL would better serve China's energy "independence" than America's.

Oh, but surely a $5.4 billion infrastructure project would provide the U.S. economy a welcome boost and added tax revenues? Yes, more than $3 billion over its lifetime, according to the market analysis in the Jan. 31 [2014] Final Supplemental Environmental Impact Statement [SEIS] on the Keystone XL prepared by the U.S. Department of State. Yet before the U.S. collects taxes from refiners, resellers, and exporters, it will first spend hundreds of millions on subsidies so these companies can invest in the technologies needed to make usable fuels out of bitumen. One example: Houston-based Motiva, which operates major storage facilities and scores of Shell gas stations and is slated to receive between $680 million and $1.1 billion from U.S. taxpayers so it can deal with tar sands oil. So in the near term, economic stimulus related to Keystone XL will come *from* Washington, not be paid *to* the IRS [Internal Revenue Service].

By facilitating tar sands oil production the proposed pipeline will result in carbon emissions equivalent to 46 new coal burning power plants.

Environmental Damage

None of these arguments should particularly matter, though, as Obama has indicated that impact on the earth's climate is his pass/fail for approving the project. This has led to a ridiculous effort to prove that the pipeline itself will not lead to a great deal more carbon entering the atmosphere. That's a feint. The real question isn't how carbon-intensive the 3-foot diameter pipe is but how much carbon-polluting oil it brings to market.

Presuming the tar sands will be developed with or without the Keystone XL, [Department of] State's estimates of carbon emissions were modest in its Jan. 31 report. Even so, the report acknowledges that the project will accelerate climate change. Hence, says Susan Casey-Lefkowitz, international program director of the Natural Resources Defense Council, "President Obama now has all the information he needs to reject the pipeline."

A new report this week [March 2014], using some of the same forecast formulas, is more damning yet. It suggests that State massively understated the consequences of the Keystone XL. According to Carbon Tracker's analysis, by facilitating tar sands oil production the proposed pipeline will result in carbon emissions equivalent to 46 new coal burning power plants.

So what explains the difference between the two reports? . . . [The] short version is that Carbon Tracker assumed that the cost of extracting a single barrel of tar sands oil and getting it to a refinery is about 30 percent higher than State's estimate, which uses today's prices. The higher number reflects an expectation that as miners have to dig deeper, the costs of extraction will rise. Also, the Keystone XL facilitates lower transportation costs and higher volume, allowing for "billions of barrels of production that otherwise would not be produced." All those billions of barrels of otherwise unproduced oil release, as [astronomer] Carl Sagan might have put it, tons and tons of otherwise unproduced carbon dioxide when they burn. . . .

The State Department report "used an extraction cost that was too low, and an oil price that was too high because they assumed climate policy would fail," says Stephen Kretzmann, executive director of Oil Change International, an energy analysis and advocacy group in Washington, D.C. "If one conducts this analysis both with accurate cost information and presuming that the U.S. will indeed meet the challenge of climate change, then it is painfully obvious that Keystone XL is a disastrous project that will unleash more than a billion cars worth of annual emissions."

Asked if Obama will consider Carbon Tracker's projections alongside State's, White House assistant press secretary Matthew Lehrich directed me to his Jan. 31 statement, in which he says a decision will take into consideration "the SEIS and *other pertinent information,* comments from the public (the deadline for those is today, Mar. 7 [2014]), and views of other agency heads." Italics are mine.

China?

Given that, my best guess is China. That's the only reason I can think for why Obama hasn't pulled the plug on Keystone XL. That, or he's waiting for Harper to promise that in exchange for it, the U.S. will never again lose to Canada in ice hockey.

7

Fracking Is Turning the US into a Bigger Oil Producer than Saudi Arabia

David Usborne

David Usborne is US editor of the Independent *and the* Independent on Sunday.

The development of fracking to access oil reserves trapped in shale sands has created a glut of oil in the United States and seems poised to make the country energy independent. The prospect of energy independence means that America may be able to reassess its relationship with dictatorial regimes in the oil-producing Middle East, and may have more leverage in its dealings with Russia.

Hector Gallegos sits in the cab of his pick-up enjoying a few hours of calm. A day earlier, workers finished carting off the huge rig that had drilled three new wells beneath this small patch of south Texas farmland and he's now getting ready to prime them for production. He reckons that about three weeks from now each will be producing 1,000 to 2,000 barrels a day. "That's money!" he exclaims with a broad smile.

It's also power, and not in the combustion sense. Thanks to the success of engineers like Mr Gallegos in pushing the frontiers of hydraulic fracturing, or "fracking", to access reserves of oil trapped in shale formations, notably here in

David Usborne, "Fracking Is Turning the US Into a Bigger Oil Producer than Saudi Arabia," *Independent*, March 11, 2014. Copyright © 2014 Independent Print Ltd. All rights reserved. Reproduced with permission.

Texas and North Dakota, America is poised to displace Saudi Arabia as the world's top producer. With that could come a hobbling of OPEC and unforeseen shifts in US foreign policy.

So rapid has been the change in its energy fortunes that even some experts, as well as policy-makers in Washington, are struggling to keep up. Nor are we just talking oil. So much natural gas is being released by the shale also that for now outlandish quantities of it are simply being burned off into the atmosphere.

Even predicting future oil output isn't the precise science you'd expect. "We keep raising our forecasts, and we keep underestimating production," Lejla Alic, an analyst with the International Energy Agency noted recently. Last year US production reached 7.4 million barrels a day, an increase over 2012 of 15.3 per cent. A jump that large hasn't been seen since 1951. This year the US should produce 8.3 million barrels a day.

Take another indicator—the volumes of crude being moved by trains, often a mile long, from the shale fields to refineries and terminals. In all of 2008, train companies moved 9,500 wagons of the black stuff. Last year, 400,000 of them rumbled across America.

The newly unlocked energy [from fracking] is set to boost the US economy and grant Washington newfound leverage around the world.

How long America's shale boom will last is hard to forecast also. In Texas, which on its own is set to increase production to 4 million barrels a day this year, the drilling peak still hasn't been reached, says Mr Gallegos. But, he suggests, "in the end it's not the oil fields or the wells that will determine where all this goes. It's the politicians around the world who set the price and make the markets." Increasingly, the decisions that matter will rest with the US, as it adjusts to its new status as a glut producer.

"The United States is now poised to become an energy superpower," writes Robert D. Blackwill and Meghan O'Sullivan, in the current issue of *Foreign Affairs* magazine.

The consequences are likely to be far-reaching, notably affording Washington a chance it hasn't had since the energy crisis of the early 1970s to reassess its relationships with those countries, often ruled by unappetising despotic governments— Saudi Arabia included—on which America has had to depend for so long to feed its fossil fuel needs.

"Since 1971, when US oil production peaked, energy has been construed as a strategic liability for the country, with its ever-growing thirst for reasonably priced fossil fuels sometimes necessitating incongruous alliances and complex obligations abroad," they write. "That logic has been upended, and the newly unlocked energy is set to boost the US economy and grant Washington newfound leverage around the world."

Among the determinations Washington must make is whether to overturn a federal law, also dating back to the early '70s, that forbids US companies from exporting crude in all but a few circumstances. Full energy independence may still be many years away, but proponents of ending the ban argue it would not only further boost the US economy—fracking added 0.3 per cent to GDP growth last year—but also help America mitigate or even end OPEC's market influence and lessen Russia's leverage also.

The Russia equation has, thanks to the Crimea crisis, raced to the fore, though it is more about gas from fracking than oil. America will see captured output touch 70 billion cubic feet a day in 2014, reaching over 100 billion cubic feet per day by 2040. In the past days, Bills have been tabled in both houses of Congress demanding that the federal government speed up the granting of licences to companies to export natural gas across the Atlantic precisely to reduce the dependency of Western Europe and Ukraine on Russian supplies.

Among those wading in is House Speaker John Boehner. "The US Department of Energy's excruciatingly slow approval process amounts to a de facto ban on American natural gas exports that [Russian president] Vladimir Putin has happily exploited to finance his geopolitical goals," he said in a statement last week.

Building the infrastructure to export quantities needed to alter the energy politics of Europe will take several years. Nor will changes in America's foreign policy stance because of its newfound oil fortunes become obvious overnight.

Indeed, some experts warn against over-stating the likely effects. "The US has a lot of interest in what's going on around the world, in the Middle East and elsewhere, regardless of whether it is independent or self-sufficient in fuels." Adam Sieminski, the administrator of the Energy Information Administration in Washington told the *Financial Post*. "Those political and economic interests will remain whether we become an exporter or not."

As vivid as the gas flares in the Texas sky at night, however, is America's new-found love affair with fracking. Environmentalists warn loudly of water contamination disasters and some home owners speak of being rattled by man-made earthquakes, but there is no giving it up now. It's a whole different world to 2008, when US oil production was at a historical low and [vice-presidential candidate] Sarah Palin was drawing liberal ire declaring that "Drill, Baby Drill!" was the answer to all of America's problems. Suddenly she seems to have been right.

Energy Innovation Needed to Reduce Dependence on Foreign Oil, Save Money

Pew Charitable Trusts

The Pew Charitable Trusts is a global nonprofit public policy organization. Its mission is to improve public policy, inform the public, and invigorate civic life.

The US Department of Defense (DoD) is hugely reliant on fossil fuels for its operations, and thus can be strongly affected by increases in the cost of fuel. This makes the DoD dependent on foreign oil and foreign price manipulation. One way to decrease this dependence is through the development of alternative fuels, especially biofuels. Congress has threatened to pass legislation preventing the use of biofuels by the DoD. This is unwise and would threaten national security. In 2012, military and civilian leaders presented a letter to Congress and President Barack Obama stressing the importance of using clean energy.

The Pew Project on National Security, Energy, and Climate today [July 24, 2012] released a letter signed by more than 350 veterans, including retired generals and admirals, as well as former Armed Services Committee chairmen Sen. John Warner and Rep. Ike Skelton, urging the president and Congress to support the Pentagon's initiatives to diversify its energy sources, limit demand and lower costs. The letter stresses

Pew Charitable Trusts, "Energy Innovation Seen As Needed to Reduce Dependence on Foreign Oil, Save Money," Pew, July 24, 2012. Copyright © 2012 Pew Charitable Trusts. All rights reserved. Reproduced with permission.

the importance of the military's ability to deploy clean energy technology to reduce dependence on fossil fuels and strengthen our national security, energy independence, and economic security.

Move to Biofuels

As the world's largest consumer of liquid fuels, the military is both becoming more energy efficient and working to test and certify advanced biofuels in its ships, planes, and vehicles. By investing in alternative fuels today, the Department of Defense (DoD) is positioning itself to take advantage of these new products when they become cost-competitive with conventional fuels. This second generation of "drop-in" biofuels is produced from domestic non-food-stock plant and biomass sources, requires no changes to current engine design, and provides the same or better performance than conventional fuels.

The development of renewable energy sources is a national security, economic, and environmental imperative.

"Today, it takes 22 gallons of fuel per soldier per day to support combat operations, a 175 percent increase over the Vietnam War era," said Phyllis Cuttino, director of the Pew Project on National Security, Energy, and Climate. "The national security community agrees that both the DoD and the nation as a whole must reduce their dependence on foreign oil. However, some in Congress are working to cripple the department's ability to move forward on energy innovation with its advanced biofuels program. This would hurt DoD's capacity to shield its budget from oil price shocks and ensure operational effectiveness."

For every $10 increase in a barrel of oil, the department pays an additional $1.4 billion annually—money that comes at the cost of operations and readiness. Some congressional amendments, if adopted, would bar DoD from purchasing or using alternative fuels and could also affect the fuels used to power unmanned vehicles for military operations.

"The bottom line is that the four branches of our military need our nation's full support to continue seeking energy solutions through innovation, as their predecessors have done for generations," added Sen. Warner, a former Navy secretary as well as former chairman of the Senate Armed Services Committee. "Our nation's energy security is linked to increasing the diversity of domestic sources of energy, both conventional and alternative, to lessen our reliance on foreign sources."

Renewable Energy and National Security

"The development of renewable energy sources is a national security, economic, and environmental imperative," Gen. Anthony Jackson, USMC [United States Marine Corps] (Ret.), said. "The next generation of Americans is deserving of our commitment to become less dependent on foreign fossil fuels."

Lt. Gen. John Castellaw, USMC (Ret.), added, "The U.S. military faces strategic, operational, and tactical vulnerabilities due to its reliance on foreign oil. Spikes in fuel costs lead to cuts in operations—reducing flying time, sailing time and training time, thereby reducing the military's overall effectiveness. We should use emerging technologies to limit these vulnerabilities."

In its report "More Fight, Less Fuel," the Defense Science Board noted: "DoD's energy problems are not unlike those of the nation. Just like the nation, to reduce its energy risks, DoD must significantly improve its energy productivity and use renewable sources where possible. . . . As these technologies find their way into commercial products, they will also limit our national dependence on foreign oil."

Energy Innovation Needed to Reduce Dependence on Foreign Oil, Save Money

Innovation has been a consistent priority and role for the U.S. military. The military's leadership, cooperation with the private sector, and early adoption have been critical to the commercialization of many technologies such as semiconductors, nuclear energy, the Internet, and the Global Positioning System. Maintaining energy innovation, inside and outside the DoD, is critical to our national security.

9

What It Would Really Take to Reverse Climate Change

Ross Koningstein and David Fork

Ross Koningstein and David Fork are engineers at Google.

Current renewable energy options are not cheap enough or convenient enough to replace fossil fuels. In order to combat climate change, therefore, entirely new technologies will need to be developed. These technologies would have to be much cheaper than fossil fuels if existing infrastructure is to be converted to the new technology. Further, to mitigate climate change, new technologies are needed to remove carbon from the atmosphere. Large scale investment in research and development is necessary if all of these new technologies are to be developed.

Google cofounder Larry Page is fond of saying that if you choose a harder problem to tackle, you'll have less competition. This business philosophy has clearly worked out well for the company and led to some remarkably successful "moon shot" projects: a translation engine that knows 80 languages, self-driving cars, and the wearable computer system Google Glass, to name just a few.

RE<C

Starting in 2007, Google committed significant resources to tackle the world's climate and energy problems. A few of these efforts proved very successful: Google deployed some of the

Ross Koningstein and David Fork, "What It Would Really Take to Reverse Climate Change," IEEE Spectrum, November 18, 2014. Copyright © 2014 IEEE Spectrum. All rights reserved. Used by permission and protected by the Copyright Laws of the United States. The printing, copying, redistribution, or retransmission of this Content without express written permission is prohibited.

most energy-efficient data centers in the world, purchased large amounts of renewable energy, and offset what remained of its carbon footprint.

Google's boldest energy move was an effort known as, RE<C which aimed to develop renewable energy sources that would generate electricity more cheaply than coal-fired power plants do. The company announced that Google would help promising technologies mature by investing in start-ups and conducting its own internal R&D [research and development]. Its aspirational goal: to produce a gigawatt of renewable power more cheaply than a coal-fired plant could, and to achieve this in years, not decades.

Unfortunately, not every Google moon shot leaves Earth's orbit. In 2011, the company decided that RE<C was not on track to meet its target and shut down the initiative. The two of us, who worked as engineers on the internal RE<C projects, were then forced to reexamine our assumptions.

To radically cut the emission of greenhouse gases, the obvious first target is the energy sector, the largest single source of global emissions.

At the start of RE<C, we had shared the attitude of many stalwart environmentalists: We felt that with steady improvements to today's renewable energy technologies, our society could stave off catastrophic climate change. We now know that to be a false hope—but that doesn't mean the planet is doomed.

New Approaches Needed

As we reflected on the project, we came to the conclusion that even if Google and others had led the way toward a wholesale adoption of renewable energy, that switch would not have resulted in significant reductions of carbon dioxide emissions. Trying to combat climate change exclusively with today's renewable energy technologies simply won't work; we need a fundamentally different approach. So we're issuing a call to action. There's hope to avert disaster if our society takes a hard look at the true scale of the problem and uses that reckoning to shape its priorities.

Climate scientists have definitively shown that the buildup of carbon dioxide in the atmosphere poses a looming danger. Whether measured in dollars or human suffering, climate change threatens to take a terrible toll on civilization over the next century. To radically cut the emission of greenhouse gases, the obvious first target is the energy sector, the largest single source of global emissions.

RE<C invested in large-scale renewable energy projects and investigated a wide range of innovative technologies, such as self-assembling wind turbine towers, drilling systems for geothermal energy, and solar thermal power systems, which capture the sun's energy as heat. For us, designing and building novel energy systems was hard but rewarding work. By 2011, however, it was clear that RE<C would not be able to deliver a technology that could compete economically with coal, and Google officially ended the initiative and shut down the related internal R&D projects. Ultimately, the two of us were given a new challenge. Alfred Spector, Google's vice president of research, asked us to reflect on the project, examine its underlying assumptions, and learn from its failures.

We had some useful data at our disposal. That same year, Google had completed a study on the impact of clean energy innovation, using the consulting firm McKinsey & Co.'s low-carbon economics tool. Our study's best-case scenario modeled our most optimistic assumptions about cost reductions in solar power, wind power, energy storage, and electric vehicles. In this scenario, the United States would cut greenhouse gas emissions dramatically: Emissions could be 55 percent below the business-as-usual projection for 2050.

Best Case, Not Good Enough

While a large emissions cut sure sounded good, this scenario still showed substantial use of natural gas in the electricity sector. That's because today's renewable energy sources are limited by suitable geography and their own intermittent power production. Wind farms, for example, make economic sense only in parts of the country with strong and steady winds. The study also showed continued fossil fuel use in transportation, agriculture, and construction. Even if our best-case scenario were achievable, we wondered: Would it really be a climate victory?

A 2008 paper by James Hansen, former director of NASA's [National Aeronautics and Space Administration] Goddard Institute for Space Studies and one of the world's foremost experts on climate change, showed the true gravity of the situation. In it, Hansen set out to determine what level of atmospheric CO_2 [carbon dioxide] society should aim for "if humanity wishes to preserve a planet similar to that on which civilization developed and to which life on Earth is adapted." His climate models showed that exceeding 350 parts per million [ppm] CO_2 in the atmosphere would likely have catastrophic effects. We've already blown past that limit. Right now, environmental monitoring shows concentrations around 400 ppm. That's particularly problematic because CO_2 remains in the atmosphere for more than a century; even if we shut down every fossil-fueled power plant today, existing CO_2 will continue to warm the planet.

We decided to combine our energy innovation study's best-case scenario results with Hansen's climate model to see whether a 55 percent emission cut by 2050 would bring the world back below that 350-ppm threshold. Our calculations revealed otherwise. Even if every renewable energy technology advanced as quickly as imagined and they were all applied globally, atmospheric CO_2 levels wouldn't just remain above 350 ppm; they would continue to rise exponentially due to continued fossil fuel use. So our best-case scenario, which was based on our most optimistic forecasts for renewable energy, would still result in severe climate change, with all its dire consequences: shifting climatic zones, freshwater shortages, eroding coasts, and ocean acidification, among others. Our reckoning showed that reversing the trend would require both radical technological advances in cheap zero-carbon energy, as well as a method of extracting CO_2 from the atmosphere and sequestering the carbon.

Although the electricity from a giant coal plant is physically indistinguishable from the electricity from a rooftop solar panel, the value *of generated electricity varies.*

Those calculations cast our work at Google's RE<C program in a sobering new light. Suppose for a moment that it had achieved the most extraordinary success possible, and that we had found cheap renewable energy technologies that could gradually replace all the world's coal plants—a situation roughly equivalent to the energy innovation study's best-case scenario. Even if that dream had come to pass, it *still* wouldn't have solved climate change. This realization was frankly shocking: Not only had RE<C failed to reach its goal of creating energy cheaper than coal, but that goal had not been ambitious enough to reverse climate change.

More Ambitious Solutions

That realization prompted us to reconsider the economics of energy. What's needed, we concluded, are reliable zero-carbon energy sources so cheap that the operators of power plants and industrial facilities alike have an economic rationale for switching over soon—say, within the next 40 years. Let's face it, businesses won't make sacrifices and pay more for clean energy based on altruism alone. Instead, we need solutions that appeal to their profit motives. RE<C's stated goal was to make renewable energy cheaper than coal, but clearly that wouldn't have been sufficient to spur a complete infrastructure change-over. So what price should we be aiming for?

Consider an average U.S. coal or natural gas plant that has been in service for decades; its cost of electricity generation is about 4 to 6 U.S. cents per kilowatt-hour [kWh]. Now imagine what it would take for the utility company that owns that plant to decide to shutter it and build a replacement plant using a zero-carbon energy source. The owner would have to factor in the capital investment for construction and continued costs of operation and maintenance—and still make a profit while generating electricity for less than $0.04/kWh to $0.06/kWh.

That's a tough target to meet. But that's not the whole story. Although the electricity from a giant coal plant is physically indistinguishable from the electricity from a rooftop solar panel, the *value* of generated electricity varies. In the marketplace, utility companies pay different prices for electricity, depending on how easily it can be supplied to reliably meet local demand.

"Dispatchable" power, which can be ramped up and down quickly, fetches the highest market price. Distributed power, generated close to the electricity meter, can also be worth more, as it avoids the costs and losses associated with transmission and distribution. Residential customers in the contiguous United States pay from $0.09/kWh to $0.20/kWh, a significant portion of which pays for transmission and distribution costs. And here we see an opportunity for change. A distributed, dispatchable power source could prompt a switchover if it could undercut those end-user prices, selling electricity for less than $0.09/kWh to $0.20/kWh in local marketplaces. At such prices, the zero-carbon system would simply be the thrifty choice.

Distributed, Dispatchable Power

Unfortunately, most of today's clean generation sources can't provide power that is both distributed and dispatchable. Solar panels, for example, can be put on every rooftop but can't provide power if the sun isn't shining. Yet if we invented a distributed, dispatchable power technology, it could transform the energy marketplace and the roles played by utilities and their customers. Smaller players could generate not only electricity but also profit, buying and selling energy locally from one another at real-time prices. Small operators, with far less infrastructure than a utility company and far more derring-do, might experiment more freely and come up with valuable innovations more quickly.

Similarly, we need competitive energy sources to power industrial facilities, such as fertilizer plants and cement manufacturers. A cement company simply won't try some new technology to heat its kilns unless it's going to save money and boost profits. Across the board, we need solutions that don't require subsidies or government regulations that penalize fossil fuel usage. Of course, anything that makes fossil fuels more expensive, whether it's pollution limits or an outright tax on carbon emissions, helps competing energy technologies locally. But industry can simply move manufacturing (and emissions) somewhere else. So rather than depend on politicians' high ideals to drive change, it's a safer bet to rely on businesses' self interest: in other words, the bottom line.

In the electricity sector, that bottom line comes down to the difference between the cost of generating electricity and its price. In the United States alone, we're aiming to replace about 1 terawatt of generation infrastructure over the next 40 years. This won't happen without a breakthrough energy technology that has a high profit margin. Subsidies may help at first, but only private sector involvement, with eager money-making investors, will lead to rapid adoption of a new technology. Each year's profits must be sufficient to keep investors happy while also financing the next year's capital investments. With exponential growth in deployment, businesses could be replacing 30 gigawatts of installed capacity annually by 2040.

Our society needs to fund scientists and engineers to propose and test new ideas, fail quickly, and share what they learn.

Remove CO_2

While this energy revolution is taking place, another field needs to progress as well. As Hansen has shown, if all power plants and industrial facilities switch over to zero-carbon energy sources right now, we'll still be left with a ruinous amount of CO_2 in the atmosphere. It would take centuries for atmospheric levels to return to normal, which means centuries of warming and instability. To bring levels down below the safety threshold, Hansen's models show that we must not only cease emitting CO_2 as soon as possible but also actively remove the gas from the air and store the carbon in a stable form. Hansen suggests reforestation as a carbon sink. We're all for more trees, and we also exhort scientists and engineers to seek disruptive technologies in carbon storage.

Incremental improvements to existing technologies aren't enough; we need something truly disruptive to reverse climate change. What, then, is the energy technology that can meet the challenging cost targets? How will we remove CO_2 from the air? We don't have the answers. Those technologies haven't been invented yet. However, we have a suggestion for how to foster innovation in the energy sector and allow for those breakthrough inventions.

Consider Google's approach to innovation, which is summed up in the 70-20-10 rule espoused by executive chairman Eric Schmidt. The approach suggests that 70 percent of employee time be spent working on core business tasks, 20 percent on side projects related to core business, and the final 10 percent on strange new ideas that have the potential to be truly disruptive.

Spur Innovation

Wouldn't it be great if governments and energy companies adopted a similar approach in their technology R&D investments? The result could be energy innovation at Google speed. Adopting the 70-20-10 rubric could lead to a portfolio of projects. The bulk of R&D resources could go to existing energy technologies that industry knows how to build and profitably deploy. These technologies probably won't save us, but they can reduce the scale of the problem that needs fixing. The next 20 percent could be dedicated to cutting-edge technologies that are on the path to economic viability. Most crucially, the final 10 percent could be dedicated to ideas that may seem crazy but might have huge impact. Our society needs to fund scientists and engineers to propose and test new ideas, fail quickly, and share what they learn. Today, the energy innovation cycle is measured in decades, in large part because so little money is spent on critical types of R&D.

We're not trying to predict the winning technology here, but its cost needs to be vastly lower than that of fossil energy systems. For one thing, a disruptive electricity generation system probably wouldn't boil water to spin a conventional steam turbine. These processes add capital and operating expenses, and it's hard to imagine how a new energy technology could perform them a lot more cheaply than an existing coal-fired power plant already does.

A disruptive fusion technology, for example, might skip the steam and produce high-energy charged particles that can be converted directly into electricity. For industrial facilities, maybe a cheaply synthesized form of methane could replace conventional natural gas. Or perhaps a technology would change the economic rules of the game by producing not just electricity but also fertilizer, fuel, or desalinated water. In carbon storage, bioengineers might create special-purpose crops to pull CO_2 out of the air and stash the carbon in the soil. There are, no doubt, all manner of unpredictable inventions that are possible, and many ways to bring our CO_2 levels down to Hansen's safety threshold if imagination, science, and engineering run wild.

The Disaster Can Be Averted

We're glad that Google tried something ambitious with the RE<C initiative, and we're proud to have been part of the project. But with 20/20 hindsight, we see that it didn't go far enough, and that truly disruptive technologies are what our planet needs. To reverse climate change, our society requires something beyond today's renewable energy technologies. Fortunately, new discoveries are changing the way we think about physics, nanotechnology, and biology all the time. While humanity is currently on a trajectory to severe climate change, this disaster can be averted if researchers aim for goals that seem nearly impossible.

We're hopeful, because sometimes engineers and scientists do achieve the impossible. Consider the space program, which required outlandish inventions for the rockets that brought astronauts to the moon. MIT [Massachusetts Institute of Technology] engineers constructed the lightweight and compact Apollo Guidance Computer, for example, using some of the first integrated circuits, and did this in the vacuum-tube era when computers filled rooms. Their achievements pushed computer science forward and helped create today's wonderful wired world. Now, R&D dollars must go to inventors who are tackling the daunting energy challenge so they can boldly try out their crazy ideas. We can't yet imagine which of these technologies will ultimately work and usher in a new era of prosperity—but the people of this prosperous future won't be able to imagine how we lived without them.

10

Ethanol Can Help to Achieve Energy Independence

National Corn Growers Association

The National Corn Growers Association (NCGA) is an association of corn growers nationwide that works to increase opportunities for farmers and expand markets for corn and corn products in the United States and abroad.

Ethanol produced from corn has numerous advantages over oil or petroleum. Ethanol is easy to renew—you simply grow more corn—whereas petroleum reserves are limited. Ethanol is less environmentally harmful to produce and manufacture than petroleum is, and the use of ethanol contributes less to greenhouse gas emissions than petroleum does. Ethanol also reduces the need for foreign reserves of oil. Ethanol is therefore important to US efforts to achieve energy independence.

With debates over ethanol heating up on the Hill [that is, in Congress] again, the National Corn Growers Association offers a comparison of the environmental impacts of ethanol and petroleum as transportation fuels. Scientifically examining a wide array of environmental factors, this side-by-side evaluation offers insight into the important differences between these fuels.

Renewable Ethanol

Which fuel is renewable and why is that important?

- Today, ethanol is primarily made from corn, which is produced annually and thereby renewable.

National Corn Growers Association, "Ethanol Offers Growing Environmental Benefits," deltafarmpress.com, April 10, 2013. Copyright © 2013 Penton Media. All rights reserved. Reproduced with permission.

When corn grows, it takes carbon dioxide from the air and converts it into glucose and then starch, from which ethanol is produced. Corn production returns nutrients to the soil through its roots and decomposing stalks, thus giving back to the land used in its production.

- Petroleum and natural gas were made over millions of years ago from decayed plants and animals. The amount present in the earth is limited, and it cannot be replenished.

As it takes tens of thousands of years for the planet to create more petroleum, it simply runs out once the current supply is exhausted. Once removed, the molecules containing carbon and other substances, like sulfur, are released into the environment. Unlike ethanol, petroleum is simply a product extracted from the planet and not one that gives back a valuable resource.

From a scientific standpoint, what actually makes up these fuels? What affect do they have on life prior to being used as fuel?

- Ethanol is a tiny single substance that is non-toxic. It can be enjoyed by adults in alcoholic beverages or as a transportation fuel.

- Petroleum is a mixture of hundreds of different molecules. It is toxic to biological organisms.

Most corn-to-ethanol production facilities are located within 15 miles of the farms where the crop was produced.

Environmental Impact

How does production of the fuel's feedstock impact the environment and the global population?

- Corn used for ethanol in the United States is grown on approximately five percent of our nation's cropland. For perspective, ethanol production uses less than three percent of all grain crops grown over the entire world.

- Petroleum is mined across the entire globe and must be extracted from deep underground. In order to collect petroleum, landscape fragmentation and the generation of toxic, hazardous and potentially radioactive waste streams often occurs.

Understanding that the distance the feedstock must travel from production to where it will be made into a useable fuel requires fuel use, how does the environmental impact of ethanol and petroleum production compare? What does it really take to make the fuel?

- Most corn-to-ethanol production facilities are located within 15 miles of the farms where the crop was produced.

Yeast, similar to that used to make bread, converts the corn starch into ethanol. In addition, two co-products, corn oil and distillers grains, are produced. These co-products are used in multiple places, including biodiesel production and animal feed.

- Since petroleum extraction happens across the globe wherever deposits can be found, it must be shipped to a facility where it can be refined. Once there, the energy intensive refining process separates these various molecules into fractions; each fraction can be used for many purposes.

What sort of air and water pollution do these fuels cause?

- Since 2005, non-toxic ethanol has replaced groundwater contaminant MTBE [methyl tert-butyl ether] as the fuel ingredient used to increase octane.

- Petroleum refiners use quite a bit of energy to separate aromatic components and very high boiling fractions for the octane needed for fuel. Many of the substances produce particulate substances leading to asthma and other health-related problems.

What type of waste is produced in the manufacture of these fuels? How do they compare in respect to the greenhouse gases [GHG] emitted in the production of these fuels?

- Based on the results of scientific testing, the EPA [US Environmental Protection Agency] considers corn starch ethanol as producing 23 percent less greenhouse gas emissions compared to making and burning gasoline from petroleum. Recent evidence shows multiple ways of producing ethanol with 50 percent or less GHG compared to gasoline production.

- The U.S. oil and gas industry generates more solid and liquid waste than municipal, agricultural, mining and other sources *combined.*

Since the RFS [Renewable Fuel Standard] was first enacted how has increased ethanol use benefitted the United States?

In 2005 and again in 2007 with the enactment of the Energy Independence and Security Act, the government chose to promote the increased production of ethanol for several reasons including its lower GHG emission properties, its renewable nature and as it decreases reliance on foreign oil. Ethanol production and use is estimated to have reduced greenhouse gas emissions by 100 million metric tons in 2012. In one year, that reduction is equivalent to removing 20.2 million light duty vehicles from the highways. Draft studies estimate a cumulative reduction of over 150 million tons of carbon dioxide emissions in the United States between the enactment of the original RFS legislation in 2005 and 2012.

- Since the enactment of the original RFS in 2005, America's oil demand has decreased, and national oil import dependence has fallen from 60 percent to 45 percent.

In 2010, U.S. oil imports fell below 50 percent for the first time since 1997. Multiple factors contributed to the decrease in petroleum usage including the increased use of ethanol, the high cost of oil and increased vehicle efficiencies.

Looking at how the production of these fuels compares side-by-side, it becomes evident that ethanol is truly renewable and produced in a greener manner than its fossil fuel counterparts. Where petroleum creates reliance upon a fuel pulled from the ground and imported from abroad, ethanol improves our environment while increasing our national and energy security.

11

Ethanol Will Not Help to Achieve Energy Independence

Jill Richardson

Jill Richardson is the founder of the blog La Vida Locavore *and a member of the Organic Consumers Association policy advisory board. She is the author of* Recipe for America: Why Our Food System Is Broken and What We Can Do to Fix It.

Proponents of corn-based ethanol argue that it can replace oil as fuel, contributing to energy independence. In fact, though, ethanol is substantially more inefficient to produce than fossil fuels, and thus is uneconomical and wasteful. Its continued production is mostly because of the political clout of corn growers, who have induced Congress to subsidize ethanol. The claim that ethanol is behind high-food prices (since corn is used for fuel instead of food) is probably exaggerated. Still, ethanol is not going to replace oil, and subsidizing it is a poor use of tax dollars.

President [Barack] Obama set the goal of moving America to 80-percent clean energy by the year 2035 in the State of the Union, only days after the EPA [Environmental Protection Agency] announced a decision to allow higher levels of corn-based ethanol in more vehicles. Obama did not mention ethanol in his speech, specifically, but he did call on us to "break our dependence on foreign oil with biofuels." And just before Christmas [2010], Congress voted for a one-year extension of ethanol subsidies. Ethanol seems to get support on both sides

Jill Richardson, "Energy Independence Goes Awry: Why the Ethanol Boom May Turn Conservation Land into Corn Fields," *Alternet*, February 10, 2011. Copyright © 2011 Alternet. All rights reserved. Reproduced with permission.

of the aisle, as new House Agriculture Committee chair Frank Lucas, R-OK, has suggested removing acres of land from a conservation program in order to grow more corn for ethanol.

The Truth About Ethanol

However, some say that ethanol is only marginally better than oil, and some blame ethanol for high global food prices. So what is the truth about ethanol? Is it the miracle fuel that will lead us to energy independence?

Measured in EROI [Energy Return on Investment], natural gas, wind, hydroelectric, and even firewood all rate better than either oil or corn-based ethanol.

About 40 percent of U.S.-grown corn now goes to produce ethanol. In the last decade, ethanol production has risen dramatically, both in terms of total corn used (from 0.6 billion bushels in 2000 to 4.9 billion bushels in 2010) and as a percent of U.S. corn (from 8 percent in 2000 to 43 percent in 2010). In the same decade, U.S. farmers increased the number of acres planted in corn from 80 million in 2000–2001 to 88 million in 2010–2011. Increases in corn yields during the same period provided a 25-percent net increase in annual corn production over the course of the decade. But even with the increase in production, the price of corn soared, from around $2 per bushel at the dawn of the new millennium to $3, $4 and even $5 per bushel over the last several years.

Given America's enthusiasm for ethanol production, one might assume that ethanol is a miracle fuel, sent from heaven to lead the way to independence from oil. (One might especially believe this if he or she works on Capitol Hill, where the entire metro station was wallpapered with ethanol ads last June [2010] reminding lawmakers that "no wars have been fought over ethanol," and "ethanol money does not support dictatorships.") But sources as diverse as *Mother Earth News* and *Popular Mechanics* provide evidence to the contrary.

Corn ethanol provides 1.3 Btu in energy for every 1 Btu consumed in producing and delivering it, according to the U.S. Department of Energy's National Renewable Energy Laboratory. A paper by SUNY [State University of New York] professor Charles Hall defines this measure as EROI, Energy Return on Investment. According to Hall, a sustainable fuel should have an EROI above 5 to 1. Oil falls short of this measure with an estimated EROI of 3 to 1, if the measure accounts for the infrastructure and energy required to extract, refine and transport the end product. But ethanol, with an EROI of 1.3 to 1, falls even shorter. Measured in EROI, natural gas, wind, hydroelectric, and even firewood all rate better than either oil or corn-based ethanol.

Political Influence

If ethanol is such an inefficient fuel, why are U.S. policies so favorable to it? *Popular Mechanics* chalks it up to money. Specifically, the money that goes to commodity growers who profit from ethanol and the money—and political influence bought with that money—of giant agribusinesses like Archer Daniels Midland. That kind of power has won them (or bought them) the support of Washington heavyweights like [retired US Army General] Wes Clark and [former Speaker of the House] Newt Gingrich. Tellingly, the CEO [chief executive officer] of Growth Energy, the pro-ethanol group co-chaired by Clark, is none other than Tom Buis, the former president of the National Farmers Union, who has years of experience working on Capitol Hill. Small wonder that agriculture and farming interests favor ethanol, especially if it is responsible for the high commodity prices of the last few years.

It seems the Grocery Manufacturers Association also believes that ethanol is behind the high prices. Unlike farmers who sell corn and companies that refine it, the companies that make processed foods sold in the supermarket are buyers of commodities, and they are not pleased at all with recent high prices. They are joined by the livestock industry, which relies on commodities like corn and soy for animal feed. Last year, several livestock industry groups asked Congress to discontinue ethanol subsidies.

Central to the debate over ethanol is the notion that it is responsible for recent high commodity prices.

House Agriculture Committee chair Frank Lucas has floated the idea of reconciling commodity growers' interests with the grocery manufacturers' and livestock producers' interests by allowing some acres currently enrolled in the U.S. Department of Agriculture's Conservation Reserve Program (CRP) to be planted in corn. Presumably, by increasing the acreage planted in corn (and therefore the supply), farmers and refiners could continue to devote a large share of U.S. corn to ethanol, while buyers of corn are saved from high prices.

CRP was created in 1985 in the midst of a farm crisis. The initial goal of the program was both conservation—taking highly erodable land out of production—and a reduction of supply and hopefully, as a result, an increase in commodity prices. Farmers who enrolled their land in the program were paid to keep the enrolled land out of production. Over the years, the program has morphed into a true conservation program, no longer aiming to reduce supply and increase prices. While it sounds ludicrous to "pay farmers to not plant" (as critics of conservation programs often claim), the program allows farmers to keep environmentally sensitive lands out of production, instead serving as wildlife habitat—a valuable use of land that the market does not normally reward financially. When asked his view on reducing CRP acres to allow more corn planting, the House Agriculture Committee's ranking member, Collin Peterson, D-MN, disagreed with the idea, saying that most of the acres currently enrolled in CRP should not be under cultivation for environmental reasons.

Ethanol and High Prices

Of course, central to the debate over ethanol is the notion that it is responsible for recent high commodity prices and that decreasing ethanol production or increasing crop acreage will reduce those prices. But the Institute for Agriculture and Trade Policy (IATP) cites ethanol and other biofuels as only one factor out of many responsible for high prices. "What we're seeing is an era of extreme volatility in agriculture markets," said IATP spokesman Ben Lilliston. "We believe it is a combination of factors that hit a perfect storm." IATP cites factors affecting actual supply and demand like increased demand for biofuels, low worldwide grain reserves, drought, and other climate-related disasters, but also manmade causes such as trade deregulation and commodity speculation.

IATP recommends taking several steps to stabilize prices, including tougher regulation of commodity markets, increasing (or creating) food reserves, increasing investment in more resilient, low-carbon agriculture, and greater trade policy flexibility. Would taking acres out of CRP to grow more corn for ethanol help? "If you believe biofuels have environmental benefits, you can start to lose those when you plant on environmentally fragile, protected land," says Lilliston. "But more importantly, it's a global market for commodity crops with many factors affecting prices, so crop expansion is not really getting to the heart of the problem. Mainly, we need government to step in and help better manage supply and price."

The future of conservation programs like CRP will not be decided until the passage of the 2012 farm bill, in a year. In the meantime, Congress will need to decide whether to renew ethanol subsidies when they expire again—a move that some believe is less likely with the bipartisan calls to cut spending in D.C.

Organizations to Contact

The editors have compiled the following list of organizations concerned with the issues debated in this book. The descriptions are derived from materials provided by the organizations. All have publications or information available for interested readers. The list was compiled on the date of publication of the present volume; the information provided here may change. Be aware that many organizations take several weeks or longer to respond to inquiries, so allow as much time as possible.

American Council for an Energy-Efficient Economy (ACEEE)
529 14th St. NW, Suite 600, Washington, DC   20045
(202) 507-4000 • fax: (202) 429-2248
website: www.aceee.org

American Council for an Energy-Efficient Economy (ACEEE) actively participates in the energy debate, developing policy recommendations and documenting how energy efficiency measures can reduce energy use, air pollutant emissions, and greenhouse gas emissions while benefiting the economy. Founded in 1980 as a nonprofit organization, ACEEE publishes numerous white papers and research reports, including *Expanding the Energy Efficiency Pie.*

American Petroleum Institute (API)
1220 L St. NW, Washington, DC   20005
(202) 682-8000
website: www.api.org

The American Petroleum Institute (API) is a national trade organization that represents approximately four hundred oil producers, refiners, suppliers, pipeline operators, marine transporters, and service and supply companies. In recent years, API has grown internationally to advocate for policies that support the oil industry worldwide. It also offers research on industry trends, sponsors research ranging from economic analyses to toxicology, disseminates environmental health and safety regulations and information, and reports on recent drilling activities and technological progress. API provides certification programs for oil industry professionals, as well as safety, environmental health, and quality control training. API publishes numerous books, pamphlets, training and safety manuals, statistical research, and newsletters each year.

**Association for the Study of Peak Oil
and Gas USA (ASPO-USA)**
1725 Eye St. NW, Suite 300, Washington, DC   20006
(202) 470-4809
e-mail: info@aspousa.org
website: http://peak-oil.org

The Association for the Study of Peak Oil and Gas USA (ASPO-USA) is a nonprofit organization focused on education and research efforts to help America understand and adapt to serious issues about energy and oil depletion. ASPO-USA, founded in 2005, is the US affiliate of ASPO International which has associated organizations in more than twenty countries. The association is a network of scientists, researchers and other energy observers examining peak oil and gas issues and the economic and scientific consequences of the oil production decline. ASPO-USA offers newsletters, statistical information, academic articles, online conference coverage, and webinars at the organization's website.

Bipartisan Policy Center (BPC)
1225 Eye St. NW, Suite 1000, Washington, DC   20005
(202) 204-2400 • fax: (202) 637-9220
e-mail: bipartisaninfo@bipartisanpolicy.org
website: http://bipartisanpolicy.org

The Bipartisan Policy Center (BPC) is a nonprofit organization that works for policy solutions through analysis and dialogue in many areas. One of its areas of expertise is energy policy. It conducts research on the rapidly changing landscape of energy needs, vulnerabilities, and opportunities, with the aim of strengthening the economy, safeguarding national security, and protecting the global environment and public health. The BPC website includes posts, blogs, and reports on energy policy.

Brookings Institution

1775 Massachusetts Ave. NW, Washington, DC   20036
(202) 797-6000 • fax: (202) 797-6004
e-mail: communications@brookings.edu
website: www.brookings.edu

The Brookings Institution is a private nonprofit organization devoted to conducting independent research and developing solutions to complex domestic and international problems. The organization's goal is to provide high-quality analysis and recommendations for decisionmakers on the full range of challenges facing an increasingly interdependent world. The Brooking Institution Press publishes multiple titles and the organization's website includes posts and reports such as *Brookings Doha Energy Forum Report 2015*.

Independent Petroleum Association of America (IPAA)

1201 15th St. NW, Suite 300, Washington, DC   20005
(202) 857-4722 • fax: (202) 857-4799
website: www.ipaa.org

The Independent Petroleum Association of America (IPAA) is a national trade association representing independent oil and natural gas producers and service companies, who develop 90 percent of domestic oil and gas wells, produce 68 percent of domestic oil, and produce 82 percent of domestic natural gas. IPAA advocates and lobbies for the interests of its members with the US Congress, federal agencies, and the executive administration. It also researches and provides economic and statistical information about the exploration and development of offshore wells. IPAA publishes an official magazine, *IPAAccess*. The IPAA website also provides a broad range of reports and statistical studies covering the oil and gas industry and supply-and-demand forecasts.

Institute for Energy Research (IER)
1100 H St. NW, Suite 400, Washington, DC   20005
(202) 621-2950 • fax: (202) 637-2420
website: www.instituteforenergyresearch.org

Founded in 1989, the Institute for Energy Research (IER) is a nonprofit organization that conducts intensive research and analysis on the functions, operations, and government regulation of global energy markets. IER promotes the idea that unfettered energy markets provide the most efficient and effective solutions to today's global energy and environmental challenges, and it works to educate legislators, policymakers, and the public to the vital role offshore drilling plays in our energy future. It publishes various fact sheets and comprehensive studies on renewable and nonrenewable energy sources, the growing green economy, climate change, and offshore oil exploration and drilling opportunities. IER also maintains a blog at its website, which provides timely comments on relevant energy and legislative issues.

Union of Concerned Scientists (UCS)
Two Brattle Square, Cambridge, MA   02138
(617) 547-5552 • fax: (617) 864-9405
website: www.ucsusa.org

Founded by scientists and students at Massachusetts Institute of Technology in 1969, the Union of Concerned Scientists (UCS) is the leading science-based nonprofit working for a healthy environment and a safer world. UCS utilizes independent scientific research and citizen action "to develop innovative, practical solutions and to secure responsible changes in government policy, corporate practices, and consumer choices." UCS publishes in-depth reports on several important issues: global warming, scientific integrity, clean energy and vehicles, global security, and food and agriculture. It also publishes the *Catalyst* magazine, *Earthwise* newsletter, and *Greentips Newsletter*. Back issues are available at its website.

United States Energy Association (USEA)
1300 Pennsylvania Ave. NW, Suite 550, Mailbox 142
Washington, DC   20004
(202) 312-1230 • fax: (202) 682-1682
e-mail: jhammond@usea.org
website: www.usea.org

The United States Energy Association (USEA) is an association of public and private energy-related organizations, corpo-
rations, and government agencies that promotes the varied interests of the US energy sector by disseminating information
about energy issues. In conjunction with the US Agency for
International Development and the US Department of Energy,
USEA sponsors an energy partnership program as well as numerous policy reports and conferences dealing with global
and domestic energy issues. USEA organizes trade and educa-
tional exchange visits with other countries. It also provides information on presidential initiatives, governmental agencies,
and national service organizations.

Bibliography

Books

Robert Bryce	*Power Hungry: The Myths of "Green" Energy and the Real Fuels of the Future.* New York: PublicAffairs, 2011.
Salvatore Carollo	*Understanding Oil Prices: A Guide to What Drives the Price of Oil in Today's Markets.* West Sussex, United Kingdom: Wiley, 2011.
Blake C. Clayton	*Market Madness: A Century of Oil Panics, Crises, and Crashes.* New York: Oxford University Press, 2015.
John H. Cushman	*Keystone & Beyond: Tar Sands and the National Interest in the Era of Climate Change.* New York: InsideClimate News, 2014.
Morgan Downey	*Oil 101.* New York: Wooden Table Press, 2009.
David Farber	*Taken Hostage: The Iran Hostage Crisis and America's First Encounter with Radical Islam.* Princeton, NJ: Princeton University Press, 2005.
Ken G. Glozer	*Corn Ethanol: Who Pays? Who Benefits?* Stanford, CA: Hoover Institute Press, 2011.

Russell Gold

The Boom: How Fracking Ignited the American Energy Revolution and Changed the World. New York: Simon & Schuster, 2014.

Richard Heinberg

Snake Oil: How Fracking's False Promise of Plenty Imperils Our Future. Santa Rosa, CA: Post Carbon Institute, 2013.

Jan H. Kalicki and David L. Goldwyn, eds.

Energy and Security: Strategies for a World in Transition. Baltimore, MD: Woodrow Wilson Center Press, 2013.

Bill McKibben, ed.

The Global Warming Reader: A Century of Writing About Climate Change. New York: Penguin Group, 2012.

Francisco Parra

Oil Politics: A Modern History of Petroleum. New York: I.B. Tauris, 2009.

Andrew T. Price-Smith

Oil, Illiberalism, and War: An Analysis of Energy and US Foreign Policy. Cambridge, MA: MIT Press, 2015.

Jeremy Shere

Renewable: The World-Changing Power of Alternative Energy. New York: St. Martin's Press, 2013.

Daniel Yergin

The Quest: Energy, Security, and the Remaking of the Modern World. New York: Penguin, 2012.

Gregory Zuckerman

The Frackers: The Outrageous Inside Story of the New Billionaire Wildcatters. New York: Portfolio, 2014.

Periodicals and Internet Sources

Richard Anderson "How American Energy Independence Could Change the World," BBC, April 3, 2014. www.bbc.com.

Jonathan Chait "The Keystone Fight Is a Huge Environmentalist Mistake," *New York*, October 30, 2013.

CNN Money "U.S. Could Be Energy Independent in Four Years," April 15, 2015. www.money.cnn.com.

Coral Davenport "New Federal Rules Are Set for Fracking," *New York Times*, March 20, 2015.

Scott Dodd "Fracking's Hidden Toll on Rural America," *Salon*, July 25, 2014. www.salon.com.

Charles Homans "Energy Independence: A Short History," *Foreign Policy*, January 3, 2012.

Keith Johnson "Europe's Energy Independence Drive Goes Off the Rails," *Foreign Policy*, February 5, 2015.

Brandon Keim "The Hot New Frontier of Energy Research Is Human Behavior," *Wired*, June 9, 2014.

Clifford Krauss "In the US, a Turning Point in the Flow of Oil," *New York Times*, October 7, 2014.

Barbara Lerner   "American Energy Independence,"
National Review, January 13, 2010.

Ryan Lizza   "The President and the Pipeline,"
New Yorker, September 16, 2013.

Jake Miller   "GOP: Energy Independence Would
Relieve Ukraine, ISIS Woes," CBS
News, September 20, 2014.
www.cbsnews.com.

Brad Plumer   "The Cost of Wind and Solar Power
Keeps Dropping All Over the World,"
Vox, February 5, 2015. www.vox.com.

Robert J.   "The U.S. May Become
Samuelson   Energy-Independent After All,"
Washington Post, November 14, 2012.

Colin Schultz   "We Don't Need a Huge
Breakthrough to Make Renewable
Energy Viable—It Already Is,"
Smithsonian, August 5, 2014.

Bryan Walsh   "The Myth of Energy Independence,"
CNN, May 10, 2013. www.cnn.com.

Washington Post   "Ethanol Takes Policy Blow from the
Environmental Protection Agency,"
November 17, 2013.

Kevin D.   "The Truth About Fracking,"
Williamson   *National Review*, February 20, 2012.

Eric Worrall   "Shocker: Top Google Engineers Say
Renewable Energy 'Simply Won't
Work,'" *WUWT*, November 22, 2014.
www.wattsupwiththat.com.

Matthew Yglesias "North American Energy
Independence: Who Cares?," *Slate*,
August 23, 2012. www.slate.com.

Index

H

Hadley, Stephen J., 8
Hall, Charles, 71
Hansen, James, 57, 61
House Agriculture Committee, 72–73
Hydraulic fracturing. *See* Fracking

I

Institute for Agriculture and Trade Policy (IATP), 73
Internal Revenue Service (IRS), 43
International Energy Agency, 47
Iranian Revolution (1979), 24
Iraq, 23
Islamic State of Iraq and Syria (ISIS), 21
Israel, 29

J

Jackson, Anthony, 52
Japan, 37
Jones Act shipping, 24

K

Keystone XL Pipeline
  approval needed, 25
  environmental damage, 43–45
  introduction, 9
  jobs and, 42–43
  no benefit from, 41–45
Kissinger, Henry, 30
Koningstein, Ross, 54–63
Korin, Anne, 28–40
Krauss, Clifford, 9

L

Lieberman, Joseph, 30
Light tight oil (LTO), 25–26
Lilliston, Ben, 73
Liquified Natural Gas (LNG), 25–26
Lucas, Frank, 70, 72
Luft, Gal, 28–40

M

Mandatory Oil Import Quota Program, 33
Massachusetts Institute of Technology (MIT), 63
McKinsey & Co., 56
Methanol, 39
Methyl tert-butyl ether (MTBE), 66
Mexico, 23
Middle East terrorist groups, 21
Miller, John, 21–27
Million barrels per day (MBD), 22–24
Motiva, 43
Mutasem, Sam, 18–20

N

National Aeronautics and Space Administration (NASA), 57
National Corn Growers Association (NCGA), 64–68
National Farmers Union, 71
National Petroleum Council (NPC), 11–12
National security concerns, 52–53